YOUNG CHILDREN AND MATHEMATICS

Teaching and Learning

Dianne S. McCarthy

innovativeink PUBLISHING
A Division of Kendall Hunt

Cover image Shutterstock.com

innovativeink PUBLISHING
A Division of Kendall Hunt

www.innovativeinkpublishing.com
Send all inquiries to:
4050 Westmark Drive
Dubuque, IA 52004-1840

Print ISBN: 979-8-3197-0246-3
eBook ISBN: 979-8-3197-0247-0

Published in the United States of America

CONTENTS

CHAPTER 1

The Importance of Math in Early Childhood

Lordn/Shutterstock.com

What is early childhood education? Early childhood education is the formal and informal educational experiences that help children grow and develop cognitively, socially, and emotionally. It typically refers to the time period from birth through age 8. This encompasses preschool, kindergarten, and grades one through three in a school setting and also includes childcare settings. And, of course, much education during this time occurs in the home. Although some consider it a "cute" time it is one of the periods of life with intense growth.

Did you know that most parents and early childhood educators consider language development the most important cognitive skill to develop in early childhood education? Do you agree? They don't believe that we need math in early childhood because we don't use it in everyday life like we do language. They also consider it developmentally inappropriate if it looks like math as they experienced it in school. David Elkind (1987) said as much in his research and writings from the 1980s. Interestingly, we do use math every day. Math involves critical thinking and problem-solving that we apply to all sorts of situations we face in life. And we can help children develop math skills appropriately.

I am currently writing this in a college library where I can see students make all sorts of math-related decisions. The line at the coffee shop is long, they must decide if they have time to wait before their next class or if it will take too long. There are students printing materials which carries with it a cost. There are two students looking at the same print out, perhaps saving money on an essential handout for class. As I sit here I think about all the math decisions I have made today before I arrived in the library. Just deciding to work in the library because of the temperature in my non-air-conditioned office was one, where to park based on the likelihood of spots at this time of day was another. I am also watching the time left on my laptop battery, something I bet some of the students are doing as well. Math surrounds us and is as pervasive as language though we just don't think about it that way. Even in early childhood kids make decisions that are math related. For example, should I put on my socks or my shoes first, should I ask for more breakfast or am I finished. Play situations offer all sorts of math tasks, will the tower of blocks stand if I put this large one on top, how many plates and cups do I need to feed the stuffed animals, and how many toy cars can I carry?

There is also much language associated with math concepts. Comparing items produces language like bigger, smaller, heavier, lighter, more, less, same, to name just a few. We use language when we describe where items are located such as above, below, and next to. These concepts are also math concepts.

Math is the study of numbers, quantities, patterns, and shapes. It includes computation, data collection and interpretation, the study of properties of numbers and shapes. Math is used to document and understand change, to analyze problems and solve them, and to reason logically about any number of things. Whereas some people consider it abstract, it has roots in concrete and pictorial representations as well. Math is the foundation of other areas of study in sciences and technology. Defining math is challenging, limiting its definition to experiences of math in school is incomplete. Math is a part of everyone's lives including those of young children. The language of math is rich.

FIVE THINGS ABOUT EARLY MATH

In 2016 Doug Clements, a Kennedy Endowed Chair in Early Childhood Learning, and the Executive Director of the Marsico Institute of Early Learning and Literacy at the University of Denver's Morgridge College of Education, presented at the White House about the interesting and surprising things about early mathematics that teachers, parents, and others don't realize. His presentation leans into five areas of early mathematics that are surprising. First, children's understanding of mathematics is broad and deep. Children below five years of age know far more that we think they do. Math achievement in preschool is an accurate predictor not just of future math achievement, but also future achievement in other fields including literacy and language. Possibly because educators believe literacy is more important than math, they lack an understanding of how mathematical thinking develops and their expectations of young children's abilities in math are often lower than they should be. They underestimate children's abilities. When mathematics teaching and learning is encouraged in early childhood education, children gain not only in mathematics, but also in social emotional experiences such as self-regulation, autonomy, and other executive function skills. They also show gains in oral language. So, even if we still hold that belief that language is more important than developing math abilities, encouraging mathematics learning and its discourse is positive. What we might really conclude here is that we don't have to choose. Developing math skills is developing oral language and literacy skills. This STEM conference where Clements spoke in April of 2016 shows clearly the importance of early math. (Mooridge College of Education University of Denver, MCEatDU, April 21, 2016, VN 000355 L65 CB283 16x9 3m30s 230620M (youtube.com)

A DEEPER DIVE INTO THE IMPORTANCE OF EARLY MATH

With some consideration of the message of Clements and the importance of early math learning let's take a deeper dive into the importance of math in early childhood education. Let's consider the impact on children and our society.

It is a basic economic principle that early learning pays off for a community and a society throughout that individual's life. Becker (1964) researched and theorized about the correlations among education level, economic productivity, and health outcomes. Better education means more economic productivity and better health throughout life. This positive research continues today. The World Bank (2018) emphasizes that investment in education at the primary level can lead to significant economic returns for a community. Heckman (2011) wrote about the value of early childhood education as a social justice issue and an economic issue with far-reaching implication. Quality early childhood education, Heckman wrote in 2006, saves a community money in things like reduced incarcerations and reduced need for public assistance like welfare (Heckman, 2006). When early childhood development is positive the child becomes a healthy adult and a productive member of society. In other words, investments in early learning pay off in productive, healthy adults that move a community forward in many ways including economically. It would make sense that mathematics teaching and learning in early childhood would help children develop into productive adults and healthier ones too, therefore, we should want to help young children learn mathematics.

Why is early math so important? It builds cognitive skills a child will use throughout life, it is the foundation of more advanced math that applies to science and technology, it is a good investment for a community as those who are learned are healthy and contribute more to a community, and math builds language skills that help children to communicate and become literate. Let's not leave math until later in a child's life.

MATH ACHIEVEMENT

The United States has not done well on international assessments in mathematics. According to the Trends in International Mathematics and Science Study (TIMSS) the United States is typically in the middle of the list of countries that engage in the study. There could be many reasons for this lack of mathematical achievement, but it is safe to say that appropriate early learning in mathematics would help people perform better when taking international assessments in grades four, eight, and twelve. Watts et al. (2014) concluded that improving early math instruction would likely improve high school and later math achievement. They also suggest that rigorous math instruction in kindergarten benefits all learners even those who may show weaker math skills upon entrance to kindergarten. There is no need to simplify math instruction in kindergarten based on entry interviews.

A myth that impacts people's beliefs about mathematics is that math abilities are inherited. Although certain areas of the brain impact attention, processing, and memory that influence mathematical ability, there is no gene that guarantees a child will be good at math, nor poor at math because of lack of that math gene. Skeide et al. (2020) found that some differences in math ability can be traced to the brain's processing system. Mathematics achievement is the result of effort and experiences including mistakes and failures (Boaler, 2015). Educators can facilitate early learning experiences in mathematics building a foundation for future math learning and life. Those experiences are not limited to counting and shape identification. They are much broader than that and will be discussed in a subsequent chapter. Early math is important because it provides experiences that build networks in the brain. Those experiences develop metacognition and mathematical intuition. Every young child is a mathematician, we need to keep the interest and achievement growing.

EARLY CHILDHOOD TEACHERS AND MATHEMATICS

As a teacher educator I have heard many well-meaning future teachers discuss showing or even doing the math for children, to not "stress them out." I doubt young children feel stress at solving a math problem the way a high school student taking an exam does. Unfortunately, some early childhood educators are the ones feeling stress or experiencing math phobia. Engaging in math is how children learn math and struggle is a part of that learning process. We need to value that engagement and support children as they learn mathematics. We need to allow math to be exciting and challenging. The struggle followed by a resolution to a challenging problem is a self-esteem builder not a stressor.

I have witnessed many children engaging in mathematics tasks over my career. Children build trains of blocks trying to make them longer than the day before and sometimes creating a pattern. Children engage in play in the cooking corner, counting out plates and play food to put on each. Children create pictures of bugs identifying the three parts of the body and adding six legs. Children deal with the problems of 26 children and three packages of 15 graham crackers for snack. Math tasks abound in early childhood informally and formally through lessons. We will now look at how to build those experiences, how math develops over time and through experiences including through play, how teachers of early childhood can design learning experiences in play-based and instructional experiences, reflect on our practices, and last how we might assess authentically and appropriately to be sure we are not underestimating what children can do mathematically.

REFERENCES

Becker, G. S. (1964). *Human capital: A theoretical and empirical analysis, with special reference to education.* University of Chicago Press.

Boaler, J. (2015). Ability and mathematics: The myths and misconceptions. *Mathematics Education Research Journal, 27*(2), 125–140.

Elkind, D. (1987). *Miseducation: Preschoolers at risk.* Random House.

Heckman, J. J. (2006). *Skill formation and the economics of investing in disadvantaged children. Science, 312*(5782), 1900–1902. https://doi.org/10.1126/science.1128898

Heckman, J. J. (2011). The economics of inequality: The value of early childhood education. *American Educator, 35*(1), 31–36.

International Association for the Evaluation of Educational Achievement. (n.d.). *TIMSS 2019 international results.* TIMSS. https://timssandpirls.bc.edu/timss2019/international-results/

Morgridge College of Education, University of Denver. (2016, April 21). *VN 000355 L65 CB283 16x9 3m30s 230620M* [Video]. YouTube. https://www.youtube.com/watch?v=230620M

Skeide, M. A., Wehrmann, K., Emami, Z., Kirsten, H., Hartmann, A. M., Rujescu, D., & Legascreen Consortium. (2020). *Neurobiological origins of individual differences in mathematical ability. PLOS Biology, 18*(10), e3000871. https://doi.org/10.1371/journal.pbio.3000871

Watts, T. W., Duncan, G. J., Siegler, R. S., & Davis-Kean, P. E. (2014). What's past is prologue: Relations between early math knowledge and high school achievement. *Educational Researcher, 43*(7), 352–360.

World Bank. (2018). *The cost of not investing in early childhood development: The case for early childhood education.* World Bank. https://documents.worldbank.org/en/publication/documents-reports/documentdetail/395141530687454547

CHAPTER 1 CONTENT ORGANIZER

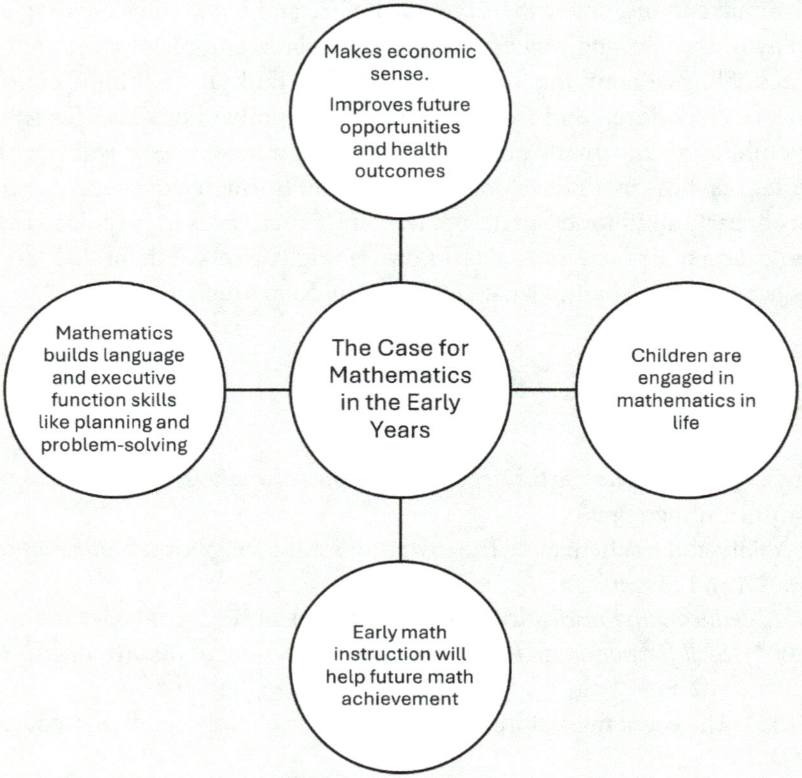

FIVE IDEAS FROM THIS CHAPTER

1. An investment in early childhood education results in returns for a community in terms of economic power and healthy outcomes.
2. We learn math as we grow and develop, engage in math experiences, and put forth effort.
3. Early math learning will help children to achieve more in future math experiences especially if they have learned that mistakes help them learn.
4. Math is pervasive in the early childhood classroom, if you just look for it.
5. Young children like to learn and do math. It is not too stressful for them.

CHAPTER 2

Essential Topics in Early Mathematics

Jacob Lund/Shutterstock.com

I remember when I started teaching kindergarten I wanted to be sure to teach math everyday and to teach it in fun and authentic ways. I also knew that math topics in early childhood are often little unconnected islands. As in today we are going to sort shapes and tomorrow we are going to skip count by fives. A well-connected web of math topics would be so much better. I decided to begin all my lessons with a problem, question, or dilemma to set a purpose and cause us to learn all sorts of math concepts and skills connected to a problem. I also used math routines like discussing a number of the day or doing a "which one doesn't belong" task. This chapter will attempt to connect those islands through a discussion of essential topics and unifying routines.

PROBLEM-SOLVING

Mathematics is a collection of many topics, yet one of them permeates all the topics. That topic is problem-solving. Problem-solving is the cause of most mathematical learning and also the outcome. Problems are why we study and learn math; we want to solve a problem. Other times we learn certain math concepts and skills and then use them to solve a problem. In either case problem-solving is most important. Appropriate mathematical problem-solving situations can arise through everyday tasks like sharing snack, situations in children's literature, games, and scientific experiences.

Schroeder and Lester (1989) coined a phrase teaching via problem-solving. This essentially says that all teaching should begin from a problem which causes the teacher to teach a math concept and sometimes a procedure. Problem-solving is, thus, the purpose of learning mathematics and the means of doing so. Problem-based learning is a similar technique where children are provided with a problem scenario where they must research and discuss to arrive at a solution. In problem-based learning with young children a scenario is presented, students discuss, ask questions, and arrive at a solution. This is an example of a scenario presented to children in second grade (McCarthy, 2001).

You are a nurse in a hospital. The doctor has ordered that a patient be given a medication in the dosage of 2.5 ml (milliliters). The liquid should be put in a dropper, but the only one you can find is say 1 ml. How should you fill it?

The students would then need to discuss and ask questions and determine that 0.5 is half of 1.0. This causes the teacher to teach decimals and fractions and probably review money because McCarthy found that students related 0.5 to 50 cents or half of a dollar.

Problem-based learning and teaching via problem-solving are somewhat in the base of the practices in *Building Thinking Classrooms* by Peter Liljedahl. Liljedahl suggests that we get kids thinking in math class. We should not just model and ask them to mimic. This is true for young children as well. They need to do age-appropriate math tasks, use their knowledge, explain their thinking, and solve problems. He suggests that young children work with partners as they problem solve. He also recommends a number of places on the internet in which to find rich problem-solving tasks. Nrich.maths.org and illuminations.nctm.org are two examples. Although some may consider the ideas of *Building Thinking Classrooms* as inappropriate for young children the philosophy of starting with problems, questions, or dilemmas is not. Allowing young children to try to solve a problem using their prior knowledge is appropriate. It is fitting to use problems to teach math concepts and skills.

When developing problem-solving abilities it is important to also consider the language of word problems. Reading and understanding the language of math and word problem is also mathematics. The Erikson Institute Early Math Collaborative has materials on a three-read strategy for helping students understand and solve word problems. It involves reading the problem three times, the first without numbers or a question, the second with numbers and quantity words and asking students to predict a question, and the third read is the full word problem that the students can then go and solve. It is important to help students understand the language of mathematics as it appears in word problems. The language of word problems should not be simplified and teachers should support children in developing an understanding of the language. Discussing the word problem as done in the three-read strategy, helping students to ask and answer their own questions to be sure they understand are essential. Drawing pictures to match word problems is another strategy that helps students understand the scenario in a word problem and then answer it. Don't rush the solution, spend time understanding the problem.

Children's literature can also provide problem-solving situations and the language of mathematics. Parents and teachers reading to and with children, asking questions, and discussing the situations presented in literature build both language and a positive disposition to mathematics learning. See a list of children's literature at the end of Chapter 3.

NUMBER SENSE

Number sense is another essential topic of early childhood mathematics education. Counting appears organically throughout young children's lives. They count cars passing by, dogs they see on a walk, the crackers on their plate, and on and on. Teachers and parents can and should encourage counting when it appears organically, yet they should also instruct in counting and number sense. Teachers, in particular, can develop experiences that help students wrestle with ideas of more, less, and same. This can be challenging when children are asked which is more 5 strawberries or 5 watermelons. Attention to magnitude or size over number may cause children to select the watermelons when these two sets are clearly the same in number. Lining up the images may help children to see that the number of each is the same even if the size of the items is not. Counting forward and backward, composing and decomposing numbers, and addition and subtraction number sentences are also topics that develop number sense.

Subitizing is the ability to quickly recognize the number of elements in a set without counting each item. An example of doing this familiar to most people is when you roll dice. You don't need to count the dots; you know the amount by the image. Subitizing is an important math skill that develops in early childhood. We can encourage its development through dice games, dominoes, and through work with sets. Grouping to subitize helps with decomposing and composing numbers. Think about rolling two dice and adding those quickly. We don't count; we know to consider the two numbers and add quickly. When looking at sets children should create minisets and then add those together. Language like I see two blocks and three marbles so that is five is what children should be striving to do.

It is also interesting to think about number sense and its development as something that is abstract. Young children do understand that 4 is more than 3 without always lining up the objects, counting, and comparing. Sometimes people think young children cannot think abstractly. Children do show elements of abstract thinking. Young children are, according to Piaget's Theory of Cognitive Development, in the pre-operational and concrete operational stages of development. Ginsburg and Opper (1988) discuss Piaget's theory in detail and conclude that age and actions are not the only part of development. Memory and recognition have roles. It is not just action on objects but reflection on the part of the child. Children need many experiences with counting and number sense. Children develop intuition about number. Intuition is often the first stage of a learning trajectory.

Working either abstractly or concretely decomposing numbers is a great way to build number sense. Baroody (2006) describes decomposition as a necessary strategy for developing fact fluency particularly for students who are neurodivergent. Cheng (2012) concluded that children aged 5–6 should use decomposition as it encourages less reliance on counting to solve addition problems. Decomposing numbers allows children to develop their own strategies for computation. A fun game for young children is go fish. You can start by matching numbers, 3 and 3, and then move into pairs that equal 10. For example, if you have a 3 in your hand you don't ask for a 3 you ask for a 7. Decomposing like this builds number sense. Other decomposition tasks are things like number bonds, number talks, and memory games where they flip cards to make a pair that equals a target number.

Last, consider doing estimation tasks to build number sense. Whether it is guessing how many candies are in a jar or estimating whether something is more or less than half help build number sense and confidence. Many people seem to think math is easy to teach because there is one answer and one strategy. There are instances in math where any of multiple strategies can be the path to an answer, sometimes multiple answers. Children need to learn that math isn't one way and one answer. They need to learn to think in ways that make sense to them. There are many resources for developing number sense on the website for the national numeracy project of the United Kingdom. See the web address at the end of the chapter. The resources are for both teachers and families.

SETS

Sets are a tool to build understanding of many mathematical concepts. Just starting with sorting and classifying objects from a set children begin to observe and discuss attributes. This is fundamental to patterning and geometry. Teachers of young children would do well to work with sets of objects, sorting, classifying, making patterns, counting them, and using them for adding and subtracting. A great game to play is to take a box of stuff or junk and then a child selects three objects as a set, the other children try to guess the attribute. Other tasks could include separating the set into two groups in different ways and then trying to separate into three or four groups. Children enjoy making collections of things. These are sets too.

SPATIAL SENSE AND GEOMETRY

Rodica Vasiliev/Shutterstock.com

Spatial sense in young children has been found to be a predictor of success in Science, Technology, Engineering and Math (STEM) fields. We need to include this topic in early childhood. It is critical that children have experiences where they can build spatial awareness and sense. This starts with understanding up/down, right/left, near/far, above/below, and other spatial concepts. Children's literature, block play, felt board play, and jigsaw puzzles do much to develop spatial sense. Building with blocks or lego or some other object also builds spatial sense. This points to the importance of all children engaging in building play. Teachers should not only encourage the play, but also engage at times, and frequently discuss with the child what they are building. Notice and wonder are too great ways to start a conversation with a child in the midst of building. For example, "I noticed that you put the small block on top of the big block, I wonder why you did that?"

Geometry and shape identification in particular are often taught in early childhood. Unfortunately, teachers often do a poor job of teaching shape identification that then has to be retaught later on. For example, teachers use things like a slice of pizza as an example of a triangle. Pizza slices that come from a round pizza are not triangles. You need straight side on a triangle. Other typical misconceptions are that squares are not rectangles, but are a unique shape. Children can create classification of items in a set and then see that a square is a special type of rectangle. We can do a much better job teaching shapes.

Measurement is another topic tied to spatial sense that we often ignore or teach poorly. Measurement systems are based on comparison. The standard length of 12 inches or 1 foot, for example, is used and related to the object you are measuring. A pencil when compared to a foot could be 1/2 foot, for example, or 6 inches. Young children do not need to learn these particular terms and numbers until they understand comparison better. There is an example from the Erikson Institute where children are asked to find something in the classroom that is the size of their hand. Now they are comparing their hand to an object. Finding something longer or shorter than their hand would be a logical next step. Comparisons such as this would be important and could lead to discussions of heavy and light, full and empty, long and short, and more and less. Even time and money fit into measurement to some extent.

LEARNING TRAJECTORIES

For more information on the development of specific math topics consider the learning trajectories developed by Clements and Sarama. See learningtrajectories.org. In most cases the trajectory for learning a specific math concept works from general intuition about the topic to understanding at an observable or concrete level, reasoning abstractly follows. When considering this it is important for early childhood educators and parents to consider how intuition develops. It comes not only from experience but also from conversations and reflections. We must talk to children about the tasks they are doing and the thinking they are engaging in. Notice and wonder, how and why, are good places to start a conversation about a math concept. How much and how often we talk with children is essential for developing their mathematical thinking and understanding. We also must allow children to grow and develop these mathematical concepts through play and high-quality instruction, not isolated direct instruction models that make math topics seem unconnected from other math topics and from the world we live in. The websites listed at the end of the chapter are good locating tasks that interesting and engaging.

REFERENCES

Baroody, A. J. (2006). Why children have difficulties mastering the basic number combinations and how to help them. *Teaching Children Mathematics*, *13*(1), 22–31.

Cheng, Z. J. (2012). Teaching young children decomposition strategies to solve addition problems: An experimental study. *The Journal of Mathematical Behavior*, *31*(1), 29–47.

Erikson Institute Early Math Collaborative. (2021, December 17). *Exploring the three-reads math protocol for word problems.* https://earlymath.erikson.edu/exploring-3-reads-math-protocol-word-problems/

Ginsburg, H.P. & Opper, S. (1988). *Piaget's Theory of Intellectual Development*. (3rd edition) Prentice Hall.

Liljedahl, P. (2021). *Building thinking classrooms in mathematics: 14 teaching practices for enhancing learning (Grades K-12).* Corwin.

McCarthy, D.S. (2001). A Teaching Experiment Using Problem-Based Learning to Develop Decimal Concepts at the Elementary Level. Doctoral Dissertation. State University of New York at Buffalo.

Schroeder, T. L., & Lester, F. K. (1989). Developing understanding in mathematics via problem solving. In P. R. Trafton & A. P. Shulte (Eds.), *New directions for elementary school mathematics* (pp. 31–42). National Council of Teachers of Mathematics.

Websites:

https://earlymath.erikson.edu
https://illuminations.nctm.org
https://www.learningtrajectories.org/
https://nationalnumeracy.org.uk
https://nrich.maths.org
https://youcubed.org

CHAPTER 2 CONTENT ORGANIZER

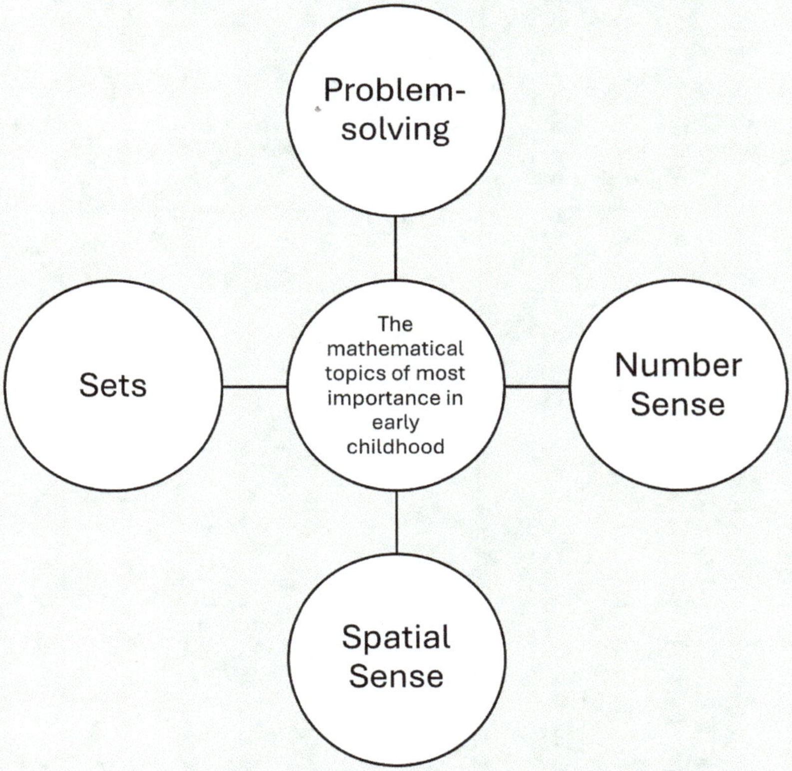

FIVE IDEAS FROM THIS CHAPTER

1. Problem-solving can be considered the cause of most mathematics teaching and learning.
2. Number sense is important to develop organically and through instruction.
3. Sets can be used to develop all sorts of mathematical understandings.
4. Spatial awareness is a predictor of future success in STEM fields. Don't ignore it.
5. Remember to consider development, learning trajectories, and the connections of mathematics topics to each other and to our lives.

CHAPTER 3

Creating an Environment for Early Math Learning

NadyaEugene/Shutterstock.com

The physical space of a classroom can be as important as the instruction a child receives in it. John Dewey explained the importance of an engaging and developmentally appropriate classroom in his writings from the early 20th century. The safety and aesthetics are important, yet the environment of the classroom should encourage exploration, risk taking, engagement, and interaction. It is extremely important to consider the environment of the early childhood classroom and how it engages the child in developing mathematical understandings and building confidence in math. It is important to also think about the home and the childcare setting when considering how the environment encourages math development. Let's consider some components of the environment as well as the space itself.

CHILDREN'S LITERATURE

Early childhood educators and parents have always considered having books as important. Having books in the home is essential for encouraging language and reading. All books have connections to math. There can be things to count, attributes to discuss, and many books discuss mathematical concepts in the plot of the story. One example would be *The Doorbell Rang* by Pat Hutchins. Children sharing cookies is the plot of the story and this clearly presents division in a realistic situation. It is important to include books in the environment, especially those with math concepts tied to the plot. See the list at the end of this chapter for some of my favorites. There are also studies and ratings of children's books for math. Check online. The Early Math Collaborative of the Erikson Institute has an organizer to help teachers analyze a book before using it. I would not refuse to use a book with a math inaccuracy. Instead discuss it with the children. Reading aloud and discussing stories should be done frequently. Books that include math concepts need to be part of the literacy program too and thus should be in the reading spaces in the classroom.

MATH MATERIALS

Just about any material can be math material, yet there are some specific toys and manipulatives that build math concepts as students play. Blocks of all sorts, shapes and sizes, pattern blocks, snap cubes, and even lego work well in classrooms so that children can build, discover principles of building, and engage in exploration of space. The ideas of next to, above, below, as well as comparing size including height can be discussed as children build. Sorting toy cars by size or creating sets from stuffed animals are ways other materials become math materials.

PLAY CENTERS AND GAMES

Play centers where children can pretend play such as kitchens, workbenches, stores with cash registers, and camping sites with gear provide opportunities for students to count, sort, and talk mathematically as they play. We need to create the play situations that cause mathematical situations to arise. The teacher should consider engaging in the play situations at times to encourage math talk.

Board games and card games should also be readily available in the early childhood classroom. Guess Who, Sorry, and Cootie are some examples of board games that build math skills along with logical reasoning and strategy. Memory, Go Fish, and card games with a standard set of 52 playing cards are great ways to build math and have fun. Jigsaw puzzles are also fun and challenging and build spatial sense. Think of all the directional words used when working with another to complete a jigsaw puzzle. A class jigsaw puzzle on a table in the room provides a great opportunity to build spatial sense and give the teacher a task for early finishers.

With just a standard set of playing cards, remove the face cards and play games with the numbers 1 through 10. Commonly called War and also called Top It in the Everyday Math program, students can play with a partner, each draws a card and the one with the higher number wins. You can do the same drawing two cards each and then adding or subtracting. It is interesting too that cards have the number of hearts or whatever shape that corresponds to the number on the card. This provides a picture representation that young children can use to learn to count in all and then to count on. For example, if a child draws an eight and a five, they can count all the shapes on the card or they can start with 8 and count on 9, 10, 11, 12, 13 to get the total. The counting on could also begin after 5 with 6, 7, 8, 9, 10, 11, 12, 13. Go Fish is a common matching game with cards. You can also play Go Fish and try to make combinations that make10. For example, if a child has a 4 in their hand they would then ask for a 6 to make 10. The internet and even AI could provide other games to be played with cards. Many of the card games also translate to dice. You can roll one or two dice, compare, add, or subtract. Kamii (2000 and 2004) found that dice games build number sense without students needing direct instruction and extensive practice with worksheets. Math for Love is a website that contains many math as play ideas and games, but even an internet search will yield suggestions for math games.

WHY PLAY IS IMPORTANT

Let's delve into why play in the early childhood classroom is important and why the play should also build math concepts and skills. Young children begin by playing alongside other children and then move into cooperative play. The dialogue as children play encourages the development of math concepts. For example, when students ask each other for a block to build with and describe it by shape and size they are using mathematical language. When they are engaged in a card game where they must compare numbers they will need to be able to explain which number is more or less. The more they play with others and use counting, sorting, describing, comparing, and explaining their reasoning the more math understandings they build and the more confident they become. Play can be a fuel for math development.

Vygtosky discusses play in his research of development. Play, particularly imaginative play, allows children to pretend and explore real life situations, think abstractly, and problem-solve. When they pretend and use objects as-if they are something else such as using a block as a phone they plan, create and follow certain rules in the play, and act like adults. This type of play allows them to develop thinking and problem-solving that are applicable to math learning.

To facilitate dialogue and discussion in a play-based experience like two students playing Top It or Go Fish, Make 10 teachers can assign a student to be referee. Flathers et al. (2024) introduced the role of referee in a second-grade classroom. The purpose of the referee was to remind students to focus, follow the rules, ask questions of players, help players, and check to make sure answers are correct. This role helps children encourage each other to explain their thinking and to focus on the math not just winning a game. It also puts a child in the role of facilitator, building autonomy and confidence.

PLACE-BASED LEARNING

Place-based learning is an instructional approach that encourages engagement with the environment, the community, and the history of the place for purposes of teaching and learning. That place for young children is the home, the classroom, and the outdoor environment of their home and school. These places provide opportunities for learning math and talking about math. Teachers can encourage discourse around what students notice. Teachers can ask questions about the size of things and the attributes. The school yard provides a great place for counting things and comparing both number and magnitude. Teachers need to look around and talk with kids about what they see whether that is in the classroom or outside. Parents should do the same in the home, yet may need encouragement from teachers that this type of talk is important. Christakis, author of *The Importance of Being Little*, says this about speaking with children and their environment.

> If I had to characterize the key difference between a high-quality and a low-quality preschool environment, it is this: in a high-quality program adults are building relationships with children and paying a lot of attention to children's thinking processes and, by extension, their communication. They attend carefully to children's language and find ways to make them think out loud. (13).

To summarize this section into easy steps the teacher and the parent or caregiver should create an environment that encourages communication through experiences and play. When I reflect on the environment in the K-3 classrooms where I taught, it was the class store, the play kitchen and restaurant, the pretend office, and the puzzles and game that accelerated math learning and made it fun. I remember in particular a store in my classroom that the students named Little Lots. Students were selling stickers, markers, and crayons.

Some of my favorite children's literature to build mathematical understandings:

Inch by Inch by Leo Leonni

Alexander Who Used to Be Rich Last Sunday by Judith Viorst

Which One Doesn't Belong by Christopher Danielson

How Many? by Christopher Danielson

Spaghetti and Meatballs for All by Marilyn Burns

Zero The Hero by Joan Holub

Equal Shmequal by Virginia Kroll

How Much Is a Million? by David M. Schwartz

A Remainder of One by Elinor Pinczes

The Button Box by Margarette S. Reid

Caps for Sale by Esphyr Slobodkina

The Doorbell Rang by Pat Hutchins

REFERENCES

Christakis, E. (2016). *The importance of being little: What young children really need from grownups.* Penguin Books

Dewey, J. (1916). *Democracy and education: An introduction to the philosophy of education.* Macmillan.

Flathers, A. M., Paternostro, L. R., Resnick, S. B, & Casa, T. M. (2024) Calling student referees to promote small-group discourse. *Mathematics Teacher: Learning and Teaching PK-12, 117*(3), 195–201.

Hynes-Berry, M. (2024). *Good Books are Good for Math.* Early Math Collaborative. https:earlymath.erikson.edu.

Kamii, C. (2000). *Young children reinvent arithmetic. Implications of Piaget's theory.* Teachers College Press.

Kamii, C. (2004) *Young children continue to reinvent arithmetic. Implications of Piaget's Theory.* Teachers College Press.

Vygtosky, L..S. (1966). Play and its role in the mental development of the child. As it appears in 2016. *International Research in Early Childhood, 7*(2). 3-25.

Websites:
https://mathforlove.com/
https://www.earlyfamilymath.org/
https://earlymath.erikson.edu

CHAPTER 3 CONTENT ORGANIZER

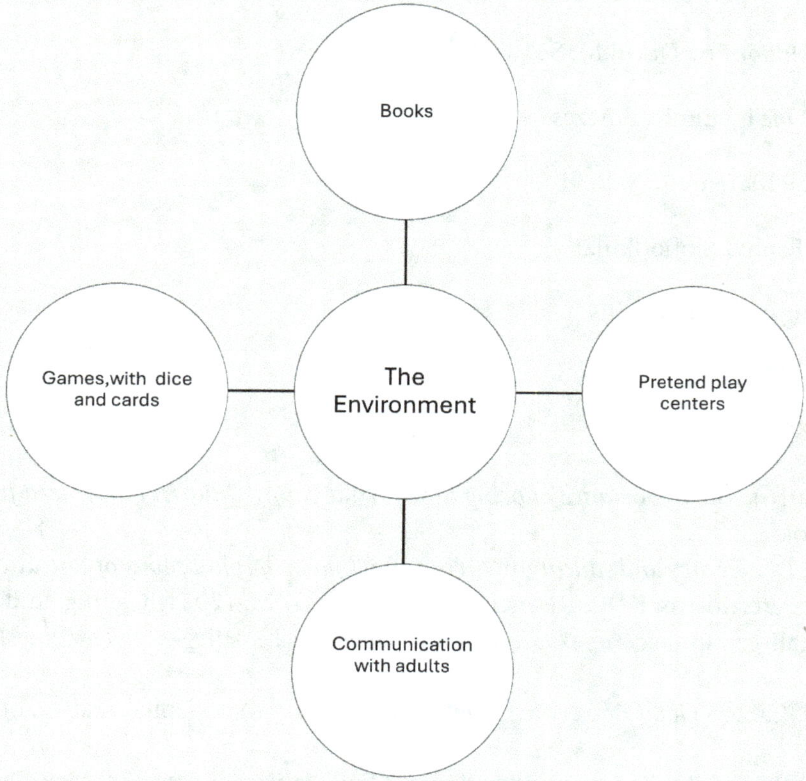

FIVE IDEAS FROM THIS CHAPTER

1. Math learning happens outside of math lessons too.
2. Communicating with children and having them explain numbers, attributes of sets, problem situations, and other mathematical situations builds mathematical reasoning.
3. Books, Games, and Play encourage math conversations and develop reasoning.
4. Children may develop their own strategies for mathematics problems.
5. Environment matters.

CHAPTER 4

Lesson Planning for Mathematics in the Early Childhood Classroom

Prostock-studio/Shutterstock.com

Most early childhood educators prefer to teach math topics when they appear organically in the classroom. For example, counting how many napkins are needed for snack time. It is possible and acceptable to plan to teach math concepts to young children. Children in grades kindergarten, 1, and 2 should receive math instruction daily. There are many decisions that a teacher can make to determine how best to teach math to young children. Lessons don't need to include worksheets and drills focused on memorizing facts and procedures. Tests need not be regular. In other words, the math lesson can be as engaging as play to learn math. Turrou (2025) writes about the narrow view early childhood educators may have about what math teaching and learning needs to look like and how these ideas lead to a phobia of school-math. Instead Turrou suggests that early childhood teachers focus on what they do well when they teach planned lessons, such as encouraging students to engage in tasks, discussing tasks with the students, and guiding the students toward a learning goal. Teachers may still teach math as it appears organically in addition to planned lessons. Planning lessons allows the teacher to think about how to help students reach a learning goal and move along the developmental trajectory. Lesson planning is a way of thinking like a teacher not a checklist. It is thinking about breaking complex ideas into smaller parts. It is about using tasks that gets learners thinking about the math concept. Lesson planning is about creating a learning experience.

LEARNING STANDARDS

All 50 states have standards for mathematics teaching and learning in grades K-2. A majority of states also have standards for preschool and prekindergarten. When considering teaching a lesson it is important to adhere to the standards and then write objectives to meet those standards. The National Association for the Education of Young Children (NAEYC) and the National Council of Teachers of Mathematics (NCTM) have standards for early learners and have joint position statements on the teaching of mathematics in the early years. Both emphasize that math instruction should be developmentally appropriate to the age and cognitive development of the child, yet rigorous enough to build number sense, problem-solving, and reasoning abilities. See their websites for the most recent statements about expectations and standards for mathematics teaching and learning for young children—www.naeyc.org and www.nctm.org.

One of the challenging skills for early childhood teachers is to analyze and understand what is expected in the standards. Let's consider a standard for kindergarten from the Common Core.

> Analyze and compare two- and three-dimensional shapes, in different sizes and orientations, using informal language to describe their similarities, differences, parts (e.g., number of sides and vertices/"corners") and other attributes (e.g., having sides of equal length).

As teachers plan instruction, they must focus not just on content but also on the action or the verbs present in the standard. In this case children are expected to analyze, compare, and describe, not just identify shapes. They must use language to show their analysis and their descriptions. Before they can begin to describe and analyze they must perceive or notice. A teacher might then consider using the questions what do you notice? How is that the same as or different from? Using these questions regularly and in many contexts will get students ready to analyze and describe in any math lesson. Teachers will then need to consider the content in the standard. What are they analyzing, comparing, and describing? For this standard they are analyzing, comparing, and describing two-dimensional and three-dimensional shapes. It is not enough to name a square or a cube, but students must explain the similarities and differences between these two shapes and compare them to other shapes. The teacher also needs to present the shapes, both 3D and 2D, in various sizes and orientations. Students should not only see but touch the shapes, and discuss what they notice about the shapes. For example, students need to think about triangles as two-dimensional shapes or flat shapes with three points or three corners or three vertices and note that a point does not have to be on the top pointing up. Now that the teacher has considered what is expected to meet this standard the lesson design can begin.

In my experience as a teacher educator, I have noticed many preservice teachers read this standard and then have students identify two-dimensional and three-dimensional shapes. That is not what the standard is expecting. Certainly, you can teach this, but it is not meeting the standard. Describing, comparing, and analyzing those shapes is critical. To meet the standard students must describe similarities and differences as they analyze and compare shapes. Same and different are concepts that are applied to every subject area, using them in math is critical.

Whether you write objectives in the traditional way—"students will be able to" statements—or more current—"I can" statements—the content and the expectations are clear. See some samples below.

> Students will be able to explain the similarities and differences among a set of two-dimensional and three-dimensional shapes.
>
> Given a set of two-dimensional and three-dimensional shapes students will analyze and sort them into groups by similar attributes.
>
> I can analyze shapes. I can explain what is similar and what is different.

After the standard and the objective have been determined the lesson plan needs to focus on what students and teachers will do. This may seem like a tedious process, yet as preservice teachers develop, lesson planning becomes a way of thinking about teaching and learning. Planning eventually gets easier and is not tedious at all. Many teachers enjoy lesson planning. If the teaching seems unsuccessful and students are not meeting the standards then planning needs to become more detailed and robust for even the most experienced teacher when a reteaching or reengagement lesson is planned.

TEACHER DECISIONS

Early childhood educators have many decisions to make when developing a lesson plan. Here is a list of either-or decisions that teachers must consider. This should be after consultation with state or common core standards and with the developmental trajectories found on the learning trajectories website. Experienced teachers may know and understand the developmental trajectory by experience. Knowing your students, their cognitive levels, interests, and motivations is also key as you consider these decisions.

- Whole group, or Small group, or Partner work for some of the time or all the time
- Manipulatives, or Pictures, or Abstract representations, or All of them
- Explicit Instruction with Teacher Modeling or Inquiry-Based Approach
- Begin with a problem situation or a story situation
- Games for practice or exercises for practice

INQUIRY-BASED APPROACH

For most math concepts in the early childhood classroom an inquiry-based approach will allow students to develop understanding in an authentic, engaging way. Inquiry-based lessons begin with a problem, question, or dilemma that is as real to life as possible, that allows for deep discussion, and that draws students into the math to be learned.

If, for example, I want to teach comparing numbers I might start with a problem scenario like this. I have two bags of candy, one bag has 21 pieces and the other has 18 pieces, which bag has more? Young children might immediately select 18 because 8 is the highest numeral but the place value is critical. An investigation can stem from this question. The teacher could provide candy or blocks or some other manipulative to illustrate the candy pieces. Students might want to draw the candy pieces on paper or a dry erase board. Students can discuss and then decide which is more and why. This type of lesson does not include teacher think alouds or modeling unless the entire class becomes completely stuck. Students should be encouraged to think and explain to their classmates. The students offer the modeling and the thinking aloud. Teachers should watch and engage with questions when necessary to guide, clarify, and extend thinking. Students can initially work with partners before small groups or whole group, or a task can be presented whole group and then continue with partners or in small groups. Again, many decisions for the teacher to consider are dependent on standards, objectives, students' abilities, students' motivation, and materials available.

Practice comparing numbers would be next in this lesson. It is important to think about how to make practice engaging. Worksheets are rarely engaging, but they do have a place and should be used at times for independent practice or assessment. For this example, I would select a game for practice. Students can be given index cards with various two-digit numbers on them and then proceed to play war or top it with a partner. (A game when each person flips up a card and the person with the higher number takes both cards.) Drawing or using cubes can help students who are unsure which number is more. A teacher can circulate, ask questions, and note who seems to grasp place value and who is struggling. Assessment does not need to be done as a paper written by a student, but can be accomplished through the observations of a teacher. Interviews also provide assessment data. The game can be the impetus for the interview with the teacher asking questions about the game and the students explaining.

If a worksheet is used it should be designed so that questions move from easier to more complex. This allows all students to engage in at least the first few examples and for some students to demonstrate their understanding of a complex example. It also needs to include space that is large enough for children to write. Too many worksheets provide small boxes that students are not able to fit what they need to write.

Closing a lesson with young children can be challenging. The teacher needs to have the students reflect on the learning and particularly what strategies they are using. Written assignments and drawing work well here as does a class-created anchor chart of how to compare numbers. Anchor charts are posters that support classroom learning by including the important content and strategies learned. They can serve as a reference to students. Creating one can also serve as formative assessment for the teacher. As students contribute the teacher realizes the content and strategies that the students have attended to the most. These can serve as assessments along with worksheets, observations, interviews, and oral answers to questions. Early childhood educators should try to use multiple means of assessment. Stories and songs also work well to consolidate the learning.

The lesson planning process allows the teacher to think about the best possible way to present content, engage students, and determine if students have met a standard. Although the lesson may not proceed as planned, the planning process helps the teacher to reflect on what to do and why. Lesson plans should not be considered checklists. When implementing a lesson plan the teacher should respond to the students' engagement, reasoning, and use of academic language. Teachers should adjust as necessary. Sometimes adjustment is going to mean throwing out the plan all together and heading down a path of learning brought about by the children. Sometimes adjustment is going to mean minor things like changing the materials used or the language. Teachers need to be experts at "reading the room". Are the children engaged? Motivated? Grasping the concepts? It is also okay to walk away from a lesson or from a math problem. In fact this practice when modeled by the teacher might help students to understand that it is okay to struggle and it is okay to take a break from a lesson and try again at another time. Champagne (2021) describes such a scenario in his first-grade class where a child asked to work on something that had become frustrating another time. The child pursued the problem the next day employing a different strategy. Champagne discussed trusting children and allowing them the freedom to try again at another time. He found it led to much success in learning. The learning in this case is not just the math, but that struggle is okay and trying other strategies and other times is also good. To be a good early childhood educator means to plan, implement, adjust, reflect, and sometimes just stop and try again another day. Teaching is not implementing a curriculum program. It is a cognitive task, and teachers need to use reasoning, reflection, observation, and critical thinking to make teaching and learning happen.

FREE RANGE TEACHING

Erika Christakis discusses in *The Importance of Being Little* the idea of planning lessons yet also allowing free range days, where students are able to pursue things of interest to them. You can also do this in your classroom building an environment where children can explore math ideas that interest them. Companies like 3M and Google are known for giving employees time to explore their own ideas or to try different solutions. Young children should be given time to explore their own ideas too.

UNIVERSAL DESIGN FOR LEARNING

Another idea to consider when planning lessons is universal design for learning (UDL). A teacher implementing the principles of universal design does so throughout the lesson planning process. Universal design is an approach to planning and teaching that is proactive, considers the needs of the students, and plans for the extremes in the classroom, such as the student that struggles the most and the one currently showing the most achievement. Universal design is about designing and teaching using multiple representations, multiple methods of assessing student learning, and multiple ways for students to engage with the content. In an article found on Edutopia Laura Moore discusses universal design in a kindergarten classroom. She describes how students show their learning in multiple ways such as stories they have written or videos they have made, how students select one of multiple locations to learn such as bean bag chairs or desks, and how students build autonomy by voting on topics for projects. All of this is done to build an inclusive classroom. Universal design is grounded in neuroscience and brain development. An in-depth description of UDL and the practices and principles that teachers should apply can be found at the website cast.org. Lambert (2021) focused on UDL and mathematics in particular. She considered specific design elements that should be applied to math classes such as making math instruction multimodal. She considered these questions when designing lessons: are students able to solve problems in their own way, is the math meaningful to the students, does the teacher interact and provide feedback that helps students to grow as mathematicians, and are students encouraged to take risks and try to solve challenging math problem? It is important that teachers apply the principles of UDL and provide that to all students in the classroom, not just those struggling, or those with individualized education plans. When using the principles of universal design math lessons become motivating and interesting to all students and they are able to engage, learn, and express their understanding of math concepts in their own ways.

Here are some templates designed for lesson planning support. It is important for teachers to find the template or planning strategy that works best for them. Because planning is about thinking like a teacher, teachers should use a template or strategy that works for them. The same is true for

assessment data and recording observations. These are some ideas, create your own or find others that work best for you.

REFERENCES

CAST. (2024). *Universal design for learning guidelines version 3.0.* https://udlguidelines.cast.org/

Champagne, Z. (2021). Walking away from a math problem is OK. *Mathematics Teacher: Learning and Teaching PK–12, 114*(9), 691–695.

Christakis, E. (2016). *The importance of being little: What young children really need from grownups.* Penguin Books.

Common Core Standards. www.corestandards.org

Lambert, R. (2021). The magic is in the margins: UDL math. *Mathematics Teacher: Learning and Teaching PK–12, 114*(9), 660–669.

Moore, L. (July 9, 2025). *Differentiated instruction: How to implement UDL in kindergarten.* Edutopia.

National Association for the Education of Young Children. (n.d.). *Position statement: Early childhood mathematics: Promoting good practices.* https://www.naeyc.org/sites/default/files/globally-shared/downloads/PDFs/resources/position-statements/psmath.pdf

National Council of Teachers of Mathematics. (n.d.). *Early childhood mathematics.* National Council of Teachers of Mathematics. https://www.nctm.org/earlychildhoodmath/

Turrou, A. C. (2025). Math phobia? Or school-math phobia? *Mathematics Teacher: Learning & Teaching PK–12, 118*(1), 80–81.

Websites

https://www.cast.org
https://www.naeyc.org
https://www.nctm.org

CHAPTER 4 CONTENT ORGANIZER

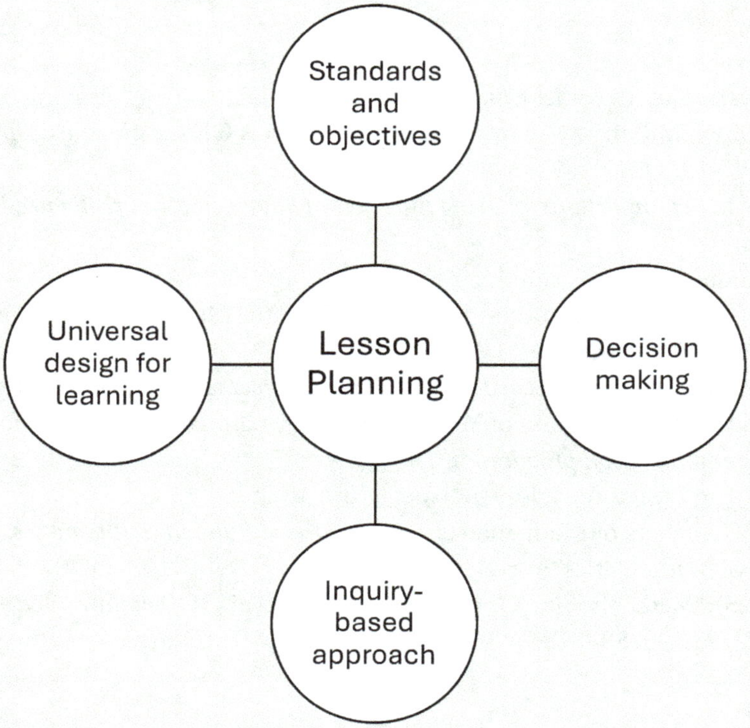

FIVE IDEAS FROM THIS CHAPTER

1. Lesson planning is a way of thinking about how to teach, not a checklist of what to do.
2. It is acceptable to plan and implement math lessons with young children; it is appropriate.
3. Even though children are young learning standards for math exist.
4. Teachers need to make many decisions as they plan.
5. Universal design principles should be applied to remove barriers and optimize learning for all students.

LESSON PLANNING TEMPLATE

Common Core or State Standards:
Learning Objective:
How the teacher will determine if the objective and the standard were met:
Anticipatory Set: Include a problem, question, or dilemma to begin. Include a list of materials needed for this step.
Procedure: List the steps and the tasks; include a list of the materials needed here.
Closure Tasks: Include a list of any materials needed.

OBSERVATION TEMPLATE

This is a way for teachers to track student achievement during a lesson and check off if a student has shown the skills and content understandings necessary.

Student Name	Lesson Objective	Standard, Verbs	Standard, Math Content	Other Information

A TEMPLATE FOR RECORDING DATA FROM A CONVERSATION WITH A CHILD

This is a way for teachers to track student achievement during a lesson and check off if a student has shown the skills and content understandings necessary.

Child's Name	Question Asked Such As....Tell me about the sets you made	Answer Child gives Such as...these all have straight sides and these have curvy sides	Child is showing.... Such as.... The ability to explain two attributes

Some good questions to ask when observing students and talking to them about their math tasks.

1. Tell me what you are doing here.
2. How did you figure that out?
3. Why did you use that strategy to....?
4. What is your answer? Why do you think that's right?
5. What does that number mean or what does that word mean?

CHAPTER 5

Lesson Study for Improving Teaching and Learning

Emese/Shutterstock.com

Although it might seem as though you are more than ready to develop and teach mathematics in your own early childhood classroom, you will never reach perfection. Things are always changing. The students and the families you work with will change, as will the content that interests and engages them. It is important for teachers to be ready for changes, embrace growth, accept that there will be failures, and work and improve as a teacher. I recall a colleague of mine who stopped a lesson that was being observed by her supervisor as evaluation of her teaching. The lesson was failing and she knew that. The teacher noticed that she needed to analyze the lesson plan, reflect, and decide what needed to change. Rather than trudge through the plan with little learning, she stopped the lesson and decided to move on to something else. This is highly unusual in an evaluative teaching situation, yet it is what an effective teacher should do. Teacher reflection has long been recommended and studied as a process for professional development. Dewey (1933) described reflection as an active and ongoing state of revisiting and revising beliefs and knowledge. Others have expanded this definition to include consideration of emotions. A strategy to build reflection into action is lesson study. In this chapter we will explore things teachers can do to continually improve their teaching and thus the learning of their students by using strategies that are related to lesson study.

LESSON STUDY

Lesson study is both an old and new strategy for teacher professional development in the United States. It began here and then fell out of favor and was adopted, refined, and expanded in Asian countries particularly Japan. It then returned to the United States and is used particularly in mathematics and language learning. More about the roots of lesson study is included in the book *Lesson Study with Mathematics and Science Preservice Teachers: Finding the Form* edited by Dotger, Matney, Heckathon, Chandler-Olcott and Fox. To summarize the process teachers work in teams study how to teach a particular topic, design a lesson, implement it, and then reflect on the process. In order to engage in lesson study there has to be reflection along with research and planning.

REFLECTION

There are two kinds of reflection to consider, reflection-in-practice, which is the thinking and decision-making that occurs while a teacher is teaching, and reflection-on-practice, which is the thinking that occurs after a lesson. Reflection-in-practice is based on formative assessment data gleaned from the students' questions and answers, completed tasks, and behaviors during the lesson. Teachers reflect-on-practice when they consider things that worked and things that did not work and how they can change things in the future. Reflection-on-practice can also consider samples of student work from the lesson. Let's delve more deeply into each of these.

Lesson planning is a way of thinking and it is future thinking as you are planning what you want to do and say to help students reach an expected outcome. Students don't always reach the expected outcome. That is okay, but as the process of teaching unveils teachers reflect on how the lesson is going and adjust to help students reach the goal. Reflection-in-practice allows teachers to change course, use different materials, or communicate in a different way to help students during the lesson. Here are some questions that teachers should ask themselves as they teach:

1. Are students moving cognitively in the direction I expected?
2. Are the answers students are giving to my questions as I expected or are they less detailed or more in-depth than I had planned?

3. Are these materials and tasks keeping students engaged? Do I see students doing what I expected with them?
4. How should I alter this lesson to help students meet the outcomes I expect right now?

Reflection-on-practice takes place after the lesson. Here the teacher looks back on the lesson through memory and through student work samples. The teacher needs to begin with the question: Did the students meet the objective I set? If not, then what needs to change in future lessons so they will meet the objective? I found in my own teaching experience that journaling helped me to reflect. I often begin with what went well and then more into what needed to change before the next lessons, whether that be the next day, or later in the week, or even when the students returned to the lesson in the same day. Journaling helped me to think actively about the changes needed. Worksheets, exit tickets, quizzes, polls, notes I made during the lesson based on what I observed, student interviews after the lesson, and my memories of the lesson provided data to back up my thoughts about the lesson.

LESSON STUDY AS COLLABORATIVE REFLECTION

Lesson study is like collaborative reflection. Lesson study has a few key components. First, there is a team of teachers working together in lesson study. The teachers do research on how best to teach a concept. This includes reading research and reviewing curriculum materials. Then the team designs a lesson utilizing the best practices for teaching this topic as well as considering the students. One teacher on the team teaches while the others observe. This is followed by reflection with student learning as the most important consideration. Did the students learn the content, did they learn it to the degree expected, and did they meet the standard? The teachers then return to the lesson plan and make changes based on the observations, the data provided by the students, and their own reflections. Lewis et al. (2009) found that lesson study changes teachers in three ways. It changes not only lesson planning, but also teachers' knowledge and beliefs and the resources they utilize. It also builds a professional community. Working with your team in a school setting and using this protocol may prove helpful in more ways than increased student learning.

Although lesson study is done collaboratively and includes a set protocol as a teacher you can utilize a similar process to help you consider the effectiveness of your teaching. Current research in the field of mathematics teaching and learning for young children can be found in publications by the National Council of Teachers of Mathematics (NCTM) and the National Association for the Education of Young Children (NAEYC). Many educators belong to these professional organizations and receive their publications. These sources are better than using materials found on Teachers Pay Teachers or Pinterest that lack review by experts in the field. See the diagram at the end of the chapter illustrating the lesson study cycle. Adapt it to fit your needs as appropriate.

STUDENT INTERVIEWS

Despite the teacher's reflections and the study of a team, students still may surprise us with their extensive abilities or lack of understanding of a math topic. To uncover this teacher should do one-on-one interviews with students. Interviews are about assessing students' understanding. They are not about teaching. Teachers need to listen, take notes, observe carefully what students do as they solve problems, and answer questions. Interviews have a positive role in improving teaching and learning because they illuminate what students are thinking and can communicate. Interviews, as described by Lewis and Fisher (2018), do not compare students to each other or against a standard in a testing situation. Instead, Lewis and Fisher explain that interview provides a context to make the child's internal reasoning external. They can illuminate difficulties as well as illustrate mathematical thinking the teacher was not aware the student was using. The following steps are helpful when designing an interview.

1. Select a standard or a concept that you wish to determine if students have met or understood.
2. Develop tasks that will help you ascertain student understanding. Consider having three tasks ready, one that is exactly what you expect the student can complete well and explain their thinking. Then one that is easier to move back, if you detect confusion, and one that is more challenging so that students can show you what they might know that you did not expect.
3. Develop a list of questions to ask that encourage the student to share their thinking.
    a. How did you get your answer?
    b. Why did you choose that strategy?
    c. What were you thinking about?
4. Implement the interview.
5. Record observations, strategies used, language used, and level of confidence and disposition.

After the interview the teacher should again reflect. This reflection is not during or after teaching, but is based on what was observed and heard in the interview. Did the student use specific language, did the student draw pictures or use manipulatives, did the student get the correct answer and is confidence evident? You, the teacher, can use this information to further develop lessons, make additions to the environment to encourage exploration and further cognitive development, and share with families.

Interviews have been shown to be an effective strategy for assessing mathematics understanding. They also improve teacher's understanding of student thinking and reasoning. Placa (2024) found that as she did interviews she better understood how children were reasoning about mathematics. She found as she focused on understanding what students were doing she asked better questions not only in the interview but also during lessons.

COLLABORATION OUTSIDE OF YOUR CLASS OR SCHOOL

Collaborating with teachers in your school situation for purposes of lesson study and reflection is important, yet there are other places where teachers can collaborate and reflect. The internet such as facebook provides groups that teachers can belong to. Here the reflection is in a virtual space, but can also be useful. Professional organizations like the NCTM have discussion forums for members. Teachers often ask for resources and help with challenges that students are facing. Collaboration in these cases comes in the form of online discussions. Although not collaborative in nature there are websites that offer resources for teachers. Sites like Edutopia include research-based articles with teaching suggestions. It is important for teachers to consider the sources and be good consumers of online materials. Information literacy needs to be practiced by all. Some good practices to consider as you peruse information online to improve your teaching and student learning are to consider the source of the information and any biases or conflicts of interest that may impact the presentation of the information. Credibility can be determined by reviewing the authors' credentials and reputation and the reputation of the publisher or website. I often tell students to triangulate information they find online. In other words, find three sources of information that agree with the idea presented in the first. Although many suggestions of how to determine the legitimacy of an idea in education point to publication dates and current practices, I do think there is a place for considering work that is older. Children today are different in many ways, but sometimes what worked before technology and before some of the practices used today are not only effective, but also better. Christakis discusses how children today can access technology and use it in unique ways such as speaking with astronauts in space, yet they often don't know the people that live next door nor are they encouraged to talk with them. This also makes me think about the benefits of using paper books and playing math games that involve cards and dice not websites. Nevertheless, consider the situation in which the author determined that a teaching practice was effective and reflected on that before moving forward with planning to use it. Information is power, lots of information about how children learn mathematics gives the teacher the power to create the best possible learning experiences.

REFERENCES

Dewey, J. (1933). *How we think, a restatement of the relation of reflective thinking to the educative process*. Heath.

Dotger, S., Matney, G., Heckathorn, J., Chandler-Olcott, K., & Fox, M. (Eds.). (2024). *Lesson study with mathematics and science preservice teachers: Finding the form*. Routledge.

Lewis, C. C., Perry, R. R. & Hurd, J. (2009). Improving mathematics instruction through lesson study: A theoretical model and North American Case. *Journal of Mathematics Teacher Education, 12,* 285–304.

Lewis, K. E., & Fisher, M. B. (2018). Clinical interviews: Assessing and designing mathematics instruction for students with disabilities. *Intervention in School and Clinic,* 53(5), 283–291.

Placa, N. (2024). Let's ask a kid!: Conducting student interviews. *Mathematics Teacher: Learning and Teaching PK-12, 117* (12), 900–906.

Websites:
https://www.edutopia.org
https://www.naeyc.org
https://www.nctm.org

CHAPTER 5 CONTENT ORGANIZER

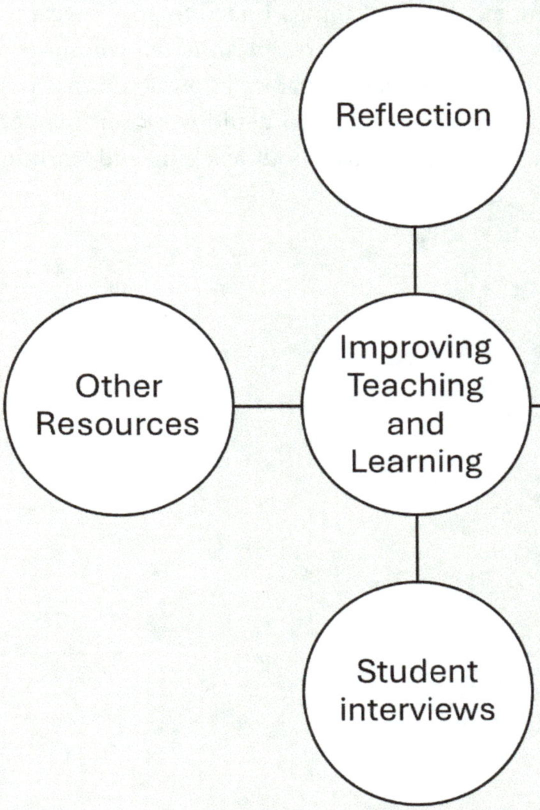

FIVE IDEAS FROM THIS CHAPTER

1. Teachers need to continually grow and develop their skills.
2. Reflection during and after teaching lessons can help improve learning.
3. Lesson study is a collaborative way of planning, teaching, and reflecting to improve student learning.
4. Student interviews give the teacher insight into students' reasoning processes and understanding.
5. There are many sources of information available to teachers seeking to improve practice.

CONCLUSION

While writing this book I would continually think of things I need to add. I wish to tell you more about writing in mathematics and using graphic organizers. I wish to tell you more about manipulatives to use for certain math topics. On the other hand what we know about mathematics teaching and learning for young children is growing and changing. I could work on this book forever and find new things to add. You too should continue to learn about teaching and learning math.

www.ingramcontent.com/pod-product-compliance
Lightning Source LLC
LaVergne TN
LVHW080527210926
848481LV00022B/3639
* 9 7 9 8 3 1 9 7 0 2 4 6 3 *